Construction Industry Council

THE CIC SCOPE OF SERVICES HANDBOOK

CIC/Handbook
first edition 2007

Guidance on use of the integrated and detailed scopes of services for use by members of the project team undertaking the definition process on major building projects

CIC Services –

Handbook

Full members of the Construction Industry Council · Association of Building Engineers · Association of Consultant Architects · Association for Consultancy and Engineering · Association for Project Management · Association for Project Safety · British Institute of Facilities Management · BRE (Building Research Establishment) · Building Services Research and Information Association · Centre for Education in the Built Environment · Chartered Institute of Architectural Technologists · Chartered Institution of Building Services Engineers · Chartered Institute of Building · Construction Industry Research and Information Association · Ground Forum · Institution of Civil Engineers · Institution of Civil Engineering Surveyors · Institute of Clerks of Works of Great Britain · Institute of Highways Incorporated Engineers · Institution of Highways & Transportation · Institute of Maintenance and Building Management · Institute of Plumbing and Heating Engineering · Institution of Structural Engineers · LABC (Local Authority Building Control) · Landscape Institute · National House-Building Council · Royal Institute of British Architects · Royal Institution of Chartered Surveyors · Royal Town Planning Institute · Steel Construction Institute · The Survey Association · **Associate members** · Association of Consultant Approved Inspectors · Association of Consultant Building Surveyors · Association of Cost Engineers · British Association of Construction Heads · British Board of Agrément · Council of Heads of the Built Environment · Chartered Institute of Marketing Construction Industry Group · Conference on Training in Architectural Conservation · Consultant Quantity Surveyors Association · Forum for the Built Environment · Federation of Property Societies · Royal School of Military Engineering · Standing Conference of Heads of Schools of Architecture · Society of Construction Law · SPONGE · Local Government Technical Advisers Group · Technology and Construction Courts' Barristers' Association · Technology and Construction Solicitors' Association

© Construction Industry Council 2007

Construction Industry Council
26 Store Street, London WC1E 7BT
tel 020 7399 7400, fax 020 7399 7425
www.cic.org.uk

RIBA ## **Publishing**

Published by RIBA Publishing
15 Bonhill Street, London EC2P 2EA
tel 020 7256 7222, fax 020 7374 2737
www.ribapublishing.com

Stock Code 61444
ISBN 978 1 85946 263 8

First published November 2007

British Library Cataloguing-in-Publication Data
A catalogue record for this book is available
from the British Library.

The publisher makes every effort to ensure the
accuracy and quality of information when it is
published. However, it can take no responsibility
for the subsequent use of this information, nor
for any errors or omissions that it may contain.

RIBA Publishing is part of RIBA Enterprises Ltd
www.ribaenterprises.com

Design by Astwood Design Consultancy
www.astwood.co.uk

Printed in Great Britain

CIC SERVICES HANDBOOK

The CIC Services Handbook

a guide to using the CIC Scope of Services

The Construction Industry Council represents the professional bodies, trade associations and research organisations in the construction industry.

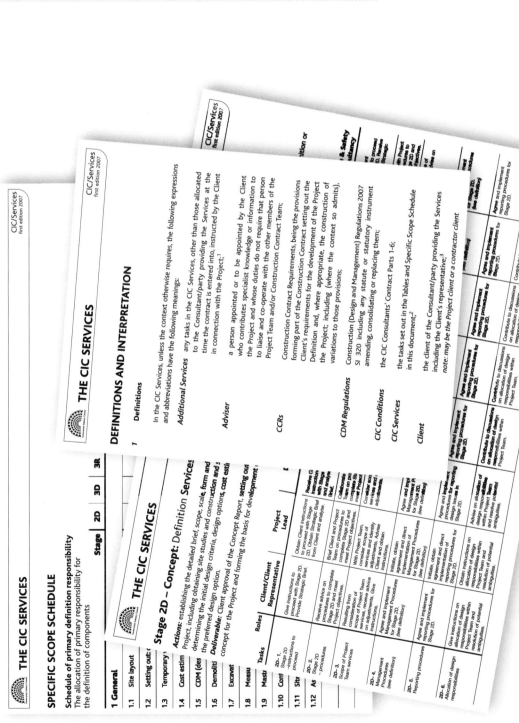

THE CIC SERVICES

SPECIFIC SCOPE SCHEDULE

Schedule of primary definition responsibility

The allocation of primary responsibility for the definition of components

	Stage	2D	3D	3R
1 General				
1.1	Site layout			
1.2	Setting out			
1.3	Temporary			
1.4	Cost estim			
1.5	CDM (des			
1.6	Demoliti			
1.7	Excavat			
1.8	Measu			
1.9	Mastr			
1.10	Con			
1.11	Site			
1.12	As			

THE CIC SERVICES

Stage 2D – Concept: Definition Services

Actions: establishing the detailed brief, scope, scale, form and determining the initial design site studies and construction and the preferred design option.

Deliverable: Client approval of the Concept Report, setting out concept for the Project and forming the basis for development

Tasks	Roles	Client/Client Representative	Project Lead
2D–1. Stage 2D –instructions to proceed		Give instructions to proceed with Stage 2D. Provide Strategic Brief	Obtain Client instructions to proceed with Stage 2D. Obtain Strategic Brief from Client and analyse.
2D–2. Stage 2D procedures		Receive advice on procedures to complete Stage 2D and meet Project objectives.	Receive C... struction Team with Stage and analyse Brief.
2D–3. Scope of Project Team services		Resulting from consideration of scope of Project Team services, receive advice on adjustments. Give instructions.	Collaborate Team to com complete Bri meet Project With Project Team, consider scope of adjustment and identify Client and obtain instructions.
2D–4. Management Procedures (see definition)		Agree and implement Management Procedures for Stage 2D. (see definition)	Initiate, obtain agreement and direct implementation of Management Procedures for Stage 2D. (see definition)
2D–5. Reporting procedures		Agree and implement reporting procedures for Stage 2D	Initiate, obtain agreement and direct implementation of reporting procedures for Stage 2D.
2D–6. Allocation of design responsibilities		Give instructions on allocation of design responsibilities within Project Team and resolution of potential	Advise on allocation of design responsibilities within Project Team and resolve potential ambiguities.

THE CIC SERVICES

DEFINITIONS AND INTERPRETATION

1   Definitions

In the CIC Services, unless the context otherwise requires, the following expressions and abbreviations have the following meanings:

Additional Services   any tasks in the CIC Services, other than those allocated to the Consultant/party providing the Services at the time the contract is entered into, instructed by the Client in connection with the Project;[1]

Adviser   a person appointed or to be appointed by the Client who contributes specialist knowledge or information to the Project and whose duties do not require that person to liaise and co-operate with the other members of the Project Team and/or Construction Contract Team;

CCRs   Construction Contract Requirements, being the provisions forming part of the Construction Contract setting out the Client's requirements for the development of the Project Definition and, where appropriate, the construction of the Project; including (where the context so admits), variations to those provisions;

CDM Regulations   Construction (Design and Management) Regulations 2007 SI 320 including any statute or statutory instrument amending, consolidating or replacing them;

CIC Conditions   the CIC Consultants' Contract Parts 1–6;

CIC Services   the tasks set out in the Tables and Specific Scope Schedule in this document;[2]

Client   the client of the Consultant/party providing the Services including the Client's representative;[3]
note: may be the Project client or a contractor client

Construction Industry Council

CIC SERVICES HANDBOOK

Overview

The CIC Scope of Services comprises ...

lists of tasks which are, or may be, required on all projects. From these lists, parties can draw up schedules of services for the appointment of consultants, specialists and contractors by allocating the tasks to whoever is to undertake them;

There are ...

fourteen Tables and the Specific Scope Schedule, together with the Definitions.

The Tables ...

set out **the services to be undertaken** by all those undertaking the definition process on major building projects;

This includes architects, engineers, quantity surveyors, CDM co-ordinators, the client, contractor and specialists.

the **Specific Scope Schedule** ...

provides a further level of detail;
see page 24

Primary definition responsibility for elements of the project (eg the façade or steelwork connections) can be allocated for each stage.

and the **Definitions** ...

set out the defined terms and abbreviations used, together with the rules of interpretation.

The services to be undertaken by **all those** ...

- which includes **Consultants**, **Specialists** and the **Contractor(s)** -
see page 12

definition + construction =

undertaking the **definition process** ...

includes **design and related activities**, such as cost, programme, health and safety, performance, integration and co-ordination.

Q: Why is it called 'the definition process'? Why not refer to 'the design'?
A: 'Definition' does not only include design; the process includes the activities of non-designers (Project Lead, Cost Consultancy and Health and Safety Consultancy) as well as those of designers.

CIC SERVICES HANDBOOK

The services are listed by **role** ...

rather than discipline, since they will be undertaken by a **range of personnel**, depending upon the contractual arrangements and responsibilities and the skills required for a particular project;
see page 8

and are **allocated** on a particular project ...

to whoever is to undertake them in an open-book approach;
see page 7

this is most easily done using **DefinIT** (the allocation software package) ...

visit www.cicservices.org.uk and follow the instructions.

The **underlying philosophy** ...

is that the activities to be undertaken to determine what is constructed, from inception to completion of a project, are fundamentally the same whatever the procurement route and whoever undertakes those activities.

Tasks	Roles	Client/Client Representative	Project Lead	Design Lead	Architectural Design	Civil & Structural Design	Building Services Design	Cost Consultancy	Health & Safety Consultancy
2D-1. Stage 2D – Instructions to ...		Give instructions to proceed with Stage 2D. Provide Strategic Brief ... instructions.	Obtain Client Instructions to proceed with Stage 2D. Obtain Strategic Brief. Submit to Client and obtain instructions.	Receive Client instructions to proceed with Stage 2D. Receive ...	Receive Client instructions to proceed with Stage 2D. Receive ... / Definition.	Receive Client instructions to proceed with Stage 2D. Receive ...	Receive Client instructions to proceed with Stage 2D. Receive ...	Receive Client instructions to proceed with Stage 2D. Receive ... / Definition.	Receive Client instructions to proceed with Stage 2D. Receive ... / Definition.
2D-8. Design approach		–	–	Establish design approach so that the design achieves required quality and h&s standard and is co-ordinated within Project Team.	With Project Team, develop design approach so that the design achieves required quality and h&s standard and is integrated within Project Team.	With Project Team, develop design approach so that the design achieves required quality and h&s standard and is integrated within Project Team.	With Project Team, develop design approach so that the design achieves required quality and h&s standard and is integrated within Project Team.	Provide cost advice to support development of design approach so that the design achieves required quality and is integrated within Project Team.	Support development of design approach so that the design achieves required quality and h&s standard and is co-ordinated within Project Team.
2D-9. Design – initial proposals		Receive report on initial proposals on design and technical viability based on Strategic Brief. Give instructions.	Based on Strategic Brief, assemble Project Team initial proposals on design and technical viability. Submit to Client and obtain instructions.	Based on Strategic Brief, co-ordinate Project Team initial proposals on design and technical viability.	Based on Strategic Brief, make initial proposals on design and technical viability.	Based on Strategic Brief, make initial proposals on design and technical viability.	Based on Strategic Brief, make initial proposals on design and technical viability.	Based on Strategic Brief, make initial proposals on Project budget viability.	Based on Strategic Brief, contribute to initial proposals on design and technical viability.
2D-10. Client priorities		Confirm Client's priorities as outlined in Strategic Brief.	Test and confirm Client's priorities as outlined in Strategic Brief.	Test and confirm Client's priorities as outlined in Strategic Brief.	Test and confirm Client's priorities as outlined in Strategic Brief.	Test and confirm Client's priorities as outlined in Strategic Brief.	Test and confirm Client's priorities as outlined in Strategic Brief.	Test and confirm Client's priorities as outlined in Strategic Brief.	–
2D-11. Property Rights		Provide information concerning Property Rights and other relevant matters, and any additional information.	Obtain from Client information concerning Property Rights and other relevant matters. Advise Client on need for additional information and obtain.	Receive information concerning Property Rights and other relevant matters. Advise on need for additional information.	Receive information concerning Property Rights and other relevant matters. Advise on need for additional information.	Receive information concerning Property Rights and other relevant matters. Advise on need for additional information.	Receive information concerning Property Rights and other relevant matters. Advise on need for additional information.	Receive information concerning Property Rights and other relevant matters. Advise on need for additional information.	Receive information concerning Property Rights and other relevant matters. Advise on need for additional information.
2D-12. H&S file		Provide any existing h&s file, Area context and ...	Receive and consider any existing h&s file ...	Receive and consider any existing h&s file ...	Receive and consider any existing h&s file ...	Receive and consider any existing h&s file ...	Receive and consider any existing h&s file ...	Receive and consider any existing h&s file ...	Receive and consider any existing h&s file and ...

CIC SERVICES HANDBOOK

Construction Industry Council

The **Tables** take an **innovative** approach ...	they look at what is usually referred to as the design process in a new way;	They are much more than the schedules of services generally found in consultants' contracts.
they are **comprehensive** ...	they set out the services to be undertaken by the whole project team from initial instructions to final completion;	
integrated and **transparent** ...	the tabular format means that it is clear what each of the participants is to do and how the tasks are inter-related and inter-dependent;	*This means that, for example, the danger of obtaining planning permission on the basis of an architectural design that is not supported by an equal level of structural or building services design becomes clearer. (The 'where do the plant rooms go?' problem.)*
they are also **flexible** ...	they can be used by a **project client** on any **procurement route** and by a **design and build contractor client**; any procurement route, includes, for example • traditional procurement (contractor tendering on full design) • design and build (both single and two stage) • develop, construct and operate • construction management and • management contracting.	*A 'project client' could be a client developing a site for its own use (an end-user) or to sell or let (a developer).*
Thus, the CIC Scope of Services provide a client with a **menu of services** ...	that will be required for a project, whether they are to be undertaken by consultants (under a professional services contract) or specialists (under other terms) and the contractor (under the construction contract). It provides a clear framework for agreeing with clients and other consultants the level of service required and the responsibilities of each party. It also provides a clear demonstration of the inter-dependence of the roles and hence the risks of allowing one role to lag behind (eg the inherent risk to an architectural concept if it has not been informed by the concept engineering).	

CIC SERVICES HANDBOOK

Application

The CIC Scope of Services can be used by Clients appointing **Consultants** ...

on the CIC Conditions (Parts 1–6 of the CIC Consultants' Contract) or on other conditions of contract;

Since the Tables and the Specific Scope Schedule include tasks to be undertaken by Consultants and others as well, the Definitions refer to 'the Consultant / the party providing the Services'.

or **Specialists** ...

that is, others (who may be specialist consultants or advisers, or specialist contractors) contributing to the definition process and appointed on other conditions of contract; *see page 12*

or **Contractors** ...

who may be engaged at an early stage to advise on buildability, cost and best value procurement, and provide advice on the definition responsibilities of the party who will be constructing the project.

The **Client** ...

is the person employing those undertaking the services.

This could be an end-user client, developer, design and build contractor or PFI project company, for example.

Q: Are the CIC Services appropriate for all projects?
A: They are written primarily for medium to large scale projects, but can be used on smaller scale projects if a full multi-disciplinary team is required.

The easiest way to prepare the schedule of services for a contract is to use DefinIT (the allocation software package). Just go to **www.cicservices.org.uk** and follow the instructions.

CIC SERVICES HANDBOOK

The Stages

There are **six stages** ...

which are:

Stage 1: Preparation

Stage 2: Concept

Stage 3: Design Development

Stage 4: Production Information

Stage 5: Manufacture, Installation and Construction Information

Stage 6: Post Practical Completion.

see page 15

Each stage has a clear start and end point and a defined level of detail, and each stage ends with a specified deliverable.

These stages reflect those adopted by many leading developers, the revised RIBA Plan of Work and the British Property Federation. They also match well with international systems.

Note: the break points between the CIC stages coincide with break points in the revised RIBA stages.

Comparison of CIC and revised RIBA Stages

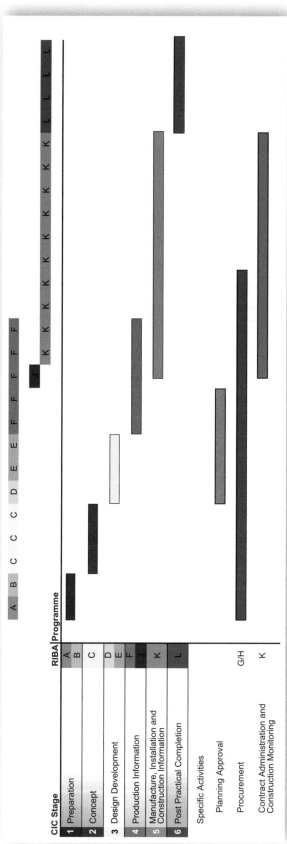

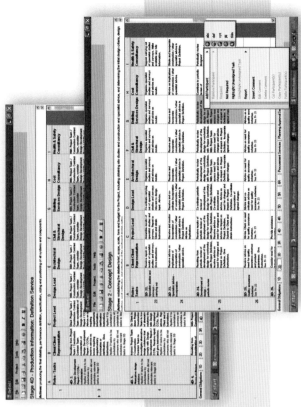

CIC SERVICES HANDBOOK

Allocation

The tasks – both in the Tables and Specific Scope Schedule – are **allocated** to whoever is to undertake them …	in so far as they are applicable for the particular project. It may well not be possible to allocate all the tasks at the outset for a variety of reasons – the procurement route may not be known and the need for other members of the project team may only emerge as the definition process proceeds, for example.

The Tables therefore determine the tasks agreed to be undertaken when the contract is formed; as the project progresses, further tasks can be allocated or tasks re-allocated, as need be.

It may not be possible to complete the whole of the Specific Scope Schedule at the outset (in particular because the range of components will not be known until the design is established), but again as the project progresses, components can be allocated or re-allocated as need be. |
| | The term 'tasks' is used to describe the activities contained in each cell of the Tables and each item in the Specific Scope Schedule.

If the client has not decided on the procurement route, it may initially only appoint its consultants for Stages 1D and 2D.

If the client has decided to appoint a design and build contractor after completion of design development (Stage 3D), it will appoint its consultants for Stages 1D, 2D and 3D but not 4D or 5D.

For most projects, nearly all the tasks in the definition stages will need to be allocated, as omitting or delaying tasks is likely to increase risk. |
The **general allocation of roles** table …	can be used to give an overall picture of the allocation of tasks. *see examples on pages 11, 18, 19 & 20*
	Visit **www.cicservices.org.uk** to find out about and access DefinIT (the allocation software package).
The Tables and Specific Scope Schedule can also be used as a **management tool** …	as they make clear the activities to be performed by each party to the contract, for all elements of a project, at any stage. By providing this framework they allow broad assumptions about roles and responsibilities to be made at the outset, to be refined as the detail of the project and procurement route becomes clear. They also give clarity to changes in scope.

Construction Industry Council

CIC SERVICES HANDBOOK

Roles

The roles ... are set out in eight columns (nine in the General Obligations and Contract Administration tables);

Roles	Client/Client Representative	Project Lead	Design Lead	Architectural Design	Civil & Structural Design	Building Services Design	Contract Administration	Cost Consultancy	Health & Safety Consultancy

First is the **Client** ...

whose role includes:
- determining the overall strategy for delivery of the project
- giving instructions and providing information
- receiving advice, information and reports from the project team
- receiving notification of decisions and instructions required and
- responding to such notifications and giving instructions and direction;

The symbols illustrated here are examples of those which can be chosen
see page 26

or **Client Representative** ...

the Client may act itself, or through a Representative identified by the Client as the Client's agent, empowered to act on the Client's behalf;

For example, a project manager may act as the Client Representative.

and **Project Lead** ...

this non-design role is one of leading and directing the project team, and includes:
- establishing the overall strategy for delivery of the project
- monitoring and integrating the activities of the project team
- developing and maintaining the programme and monitoring progress
- reporting to and seeking decisions and instruction from the client and obtaining them.

The Project Lead role may be undertaken by a project manager, for example (but not the Client Representative), or by someone with another role, such as an architect undertaking the architectural design.

Then the design roles of **Design Lead** ...

which includes:
- setting design standards and co-ordinating and integrating the design
- co-ordinating advice on design-related issues and providing advice.
On a building project, the role is likely to be undertaken by the same party as the architectural design; but it could be the same party as undertakes the civil and structural or building services design.

For example, on a data centre, the Design Lead role may be undertaken by a building services engineer.

CIC SERVICES HANDBOOK

Architectural Design ...

which covers the traditional architectural services.

Generally, up to production information (Stage 4D) the role will be undertaken by an architect or architectural technologist, or in part by a landscape architect.

At the manufacture, installation and construction information stage (Stage 5D), the architectural design tasks will generally be undertaken by the contractor under the construction contract.

In addition, some tasks may be undertaken by specialists.

The Architectural Design role may be undertaken by more than one person, for example an architect and a landscape architect may be appointed.

Civil and Structural Design ...

which covers the traditional civil and structural engineering design services.

Generally, the role will be undertaken by a consulting engineer up to production information (Stage 4D) and particular aspects may be undertaken by specialists.

At the manufacture, installation and construction information stage (Stage 5D), the tasks will generally be undertaken by the contractor under the construction contract (and the contractor will use subcontractors, specialist fabricators and suppliers).

For example, steelwork connections and pre-stressed concrete details may be designed by a specialist in Stage 4D.

---

and **Building Services Design** ...

which covers the traditional mechanical, electrical and public health design services.

In the industry, there is a wide variation in practice as to the level of detail or definition information produced by a consulting engineer; the end of concept (Stage 2D), design development (Stage 3D) or production information (Stage 4D) provide suitable points for handover to a specialist design contractor.

Different consultants may undertake mechanical, electrical and public health services.

For example, a client may appoint a consulting building services engineer for Stages 1D, 2D and 3D, and a specialist contractor for Stage 4D.

BSRIA's *A Design Framework for Building Services* uses the same stages and terminology as the CIC Services, and provides a further level of detail.

CIC SERVICES HANDBOOK

---

and the non-design roles
of **Cost Consultancy** …

which covers the traditional function of quantity surveying;

and **Health and Safety
Consultancy** …

which includes, but is not limited to, the services of the CDM co-ordinator under the Construction (Design and Management) Regulations 2007;

the person undertaking the role is a full member of the project team and takes a full part in the definition process.

NB All those undertaking definition services have an obligation to comply with health and safety legislation, in particular the CDM Regulations, and this is expressly referred to in GO—13; thus if a party is in breach of the CDM Regulations, they could be liable in damages to their client under their contract as well as liable to prosecution for breach of the legislation.

Elements of the
definition process …

the tasks to be undertaken by each participant are limited to **those elements or aspects** of the design and definition process or works which are **within the scope of the relevant role.**

For example, the work of an architect undertaking the architectural design role will be limited to the architectural aspects of the design and definition process; a quantity surveyor's work will be limited to cost and related aspects, and a health and safety co-ordinator the health and safety aspects.

ng and

The tasks to be undertaken are:
- those allocated, in so far as appropriate for the Project limited to those elements of the Project Definition or Works which are within the scope of the relevant role and
- limited to those aspects of the definition or review of the Project Definition or Works which are within the scope of the relevant role.

Architectural	Civil & Structural	Building Services	Cost	Health & Safety

**Contract
Administration**

that is, the role of contract administration under or in relation to the construction contract.

The contract administration role may for example be undertaken by an architect who is undertaking the architectural design.

An employer's agent, acting for the client under a design and build contract, may undertake some contract administration tasks.

An example of the general allocation of roles for traditional procurement follows on the next page

10

CIC SERVICES HANDBOOK

Example of traditional procurement: developer Client's Consultants

Legend:
- [diagonal slash] not used
- [grey box] to be allocated
- [grey box] inapplicable

GENERAL OBLIGATIONS	Client/Client Representative (= client)		Project Lead (= project manager)		Contract Admin (= architect)	Design Lead (= architect)		Architectural Design (= architect)		Civil & Structural Design (= structural engineer)		Building Services Design (= m&e engineer)		Cost Consultancy (= quantity surveyor)		Health & Safety Consultancy (= CDM coordinator)	
STAGES	Definition	Review	Definition	Review		Definition	Review	Definition	Review	Definition	Review	Definition	Review	Definition	Review	Definition	Review
STAGE 1 Preparation	●		●			●		●		●		●		●		●	●
STAGE 2 Concept	●		●			●		●		●		●		●		●	●
STAGE 3 Design Development	●	/	●			●	/	●	/	●	/	●		●	/	●	
STAGE 4 Production Information	●	/	●			●	/	●	/	●	/	~	~	●		●	/
STAGE 5 MI&C Information	~	●	~	●		~	●	~	●	~	●		●	~	●	~	●
STAGE 6 Post Practical Completion	●		●			●		●		●		●		●		●	
Procurement	●		●			●		●		●		●		●		●	
Planning Approval	●		●					●		●		●		●			
Contract Admin etc	●		●			●		●		●		●		●		●	

CIC SERVICES HANDBOOK

Participants

Turning to the defined terms, the Project Team ...

comprises the **Consultants** and **Specialists,** in particular those undertaking the various roles; they work together, liaise and co-operate with the Client and each other;

See the Definitions in the CIC Scope of Services for the full definitions.

Consultants ...

are those appointed by the Client to undertake a consultant role, and to whom tasks have been or are allocated;

Specialists ...

are those (other than Consultants) appointed by the Client

• who contribute specialist knowledge or information to the definition (or review of the definition) of the project or

• who carry out specialist investigations, tests or studies and

• whose duties require them to liaise and co-operate with the other members of the Project Team and

• to whom tasks may be allocated.

For example, a specialist fabric roof contractor or specialist contractor designing the facade may undertake the design role for those elements; or a planning consultant or engineer doing an environmental survey may be allocated some tasks in the Tables for specific aspects of the project (and allocated tasks in the Specific Scope Schedule).

Advisers ...

also contribute specialist knowledge or information but are not members of the Project Team.

For example, an adviser on rights of light or a party wall surveyor would be an Adviser.

Construction Contract Team ...

is the term used for the Project Team where the Client is a contractor (eg under a design and build contract).

The Contractor ...

with a capital 'C', is the person appointed by the Client under the Construction Contract to carry out or manage the construction and (if required, for example under a design and build contract) to carry out or manage the development of the definition process for the whole or part of the construction of the Project;

a contractor (with a little 'c') may be part of the Project Team during the early stages of the Project (and be described as a Specialist or Consultant).

At Stage 5D (the Manufacture, Installation and Construction Information Stage), as party to the Construction Contract, the Contractor will be involved in the continuing definition process, by defining what is to be constructed under the Construction Contract, preparing manufacturing and installation drawings (as a minimum).

The Tables

The fourteen tables are **General Obligations** …	which apply generally, throughout all stages and activities;
the six stages …	described on page 6 and in more detail on pages 15 and 16; and
three specific activities, **Procurement, Planning Approval** and **Contract Administration and Contract Monitoring**	which are not time dependent and do not therefore occur at a specific stage.

For example, planning approval will be applied for at different stages on different projects – and approval will be obtained at different times. The procurement of consultants, specialists and contractors will also occur at different times throughout the job.

services …	*If construction monitoring services are required but not contract administration, the latter tasks may be deleted.*
and **Further Services** …	which may be instructed by the client, either at the time the contract is entered into, or at a later stage.

For example, a client may ask a consultant to include in their initial fee whole life cost studies (see FS– 2) or a site survey (see FS– 12), meaning that these services become part of the contract. At a later stage, planning permission may be refused and the client may instruct the consultant to appeal (see FS– 15), meaning that an additional fee will be payable for the additional service.

Specific Activities

Procurement Services *(Pr– 1 to Pr– 45)* 57

The procurement of Consultants, Specialists and the Contractor (single stage and two stage tendering) and incidental services.

Planning Approval Services *(PA– 1 to PA– 17)* 66

The obtaining of planning approval and incidental services.

Contract Administration and Construction Monitoring Services *(CaCm– 1 to CaCm– 28)* 69

The administration of the Construction Contract and monitoring the construction of the Works on site.

Further Services *(FS– 1 to FS– 37)* 75

Further services which may be instructed by the Client.

The numbered rows in Procurement are prefixed 'Pr', in Planning Approval 'PA' and in Contract Administration and Construction Monitoring 'CaCm.' Further Services are prefixed 'FS'.

CIC SERVICES HANDBOOK

Definition and Review Services

The tasks are divided into **definition services** ...	including design, specification, cost control, programming, administration, compliance with regulations and health and safety for example;
	The definition stages are Stages 1D, 2D, 3D, 4D and 5D (that is, all stages up to practical completion).
which may be undertaken by the **project client's** team ...	under the direction and instruction of the project client;
	For example, a project client using traditional procurement may appoint consultants and specialists to undertake the definition services from preparation to production information (Stages 1D to 4D).
or a **design and build contractor's** team ...	of consultants together with specialists (who may be sub-contractors or suppliers), under the direction and instruction of the contractor client;
	Under a design and build contract, the contractor will take on responsibility for definition services from the time it is appointed, which may for example be design development or production information (Stages 3D or 4D).
and **review services** ...	where the definition output of others is reviewed for general compliance with design intent.
	For example, a developer who has appointed a design and build contractor at the end of Stage 2D may wish its consultants to review the design undertaken by the contractor's designers against the employer's requirements and contractor's proposals. Therefore the contractor will be responsible for definition services in Stages 3D, 4D and 5D and the project client's consultants will undertake Stages 3R, 4R and 5R services.
	The review services at the manufacture, installation and construction information stage (Stage 5R) will generally be undertaken by the project client's team, reviewing the work undertaken by the contractor.
	The review stages are Stages 3R, 4R, 5R and 6R.
	In design and build contracting, the project client may retain its consultants to review the design carried out by the contractor.
	In traditional procurement, Stage 5D services will be undertaken by the contractor under the construction contract, and the client's consultants will review that design, undertaking Stage 5R services.
	If production information (Stage 4D) services for a particular discipline (eg building services) or components (eg steelwork connections) are undertaken by specialist contractors, the client's consultants will often review that work, undertaking Stage 4R services.

CIC SERVICES HANDBOOK

Deliverables at each stage

Each stage ...
is headed with a summary of the actions to be undertaken and the deliverable.

Stage 1D –
Preparation ...
covers defining the Project objectives, business need, Client priorities and aspirations; describing the acceptance criteria, including function, mix of uses, scale, location, quality, cost, value, time, safety, health, environment and sustainability;

and concludes with the Client approving the Strategic Brief, setting out the Stage 1D definition information and forming the basis for the development of the concept definition.

Stage 2D –
Concept ...
covers establishing the detailed brief, scope, scale, form and budget for the Project, including obtaining site studies and construction and specialist advice, and determining the initial design criteria, design options, cost estimates and selection of the preferred design option;

and concludes with the Client approving the Concept Report, setting out the integrated concept for the Project and forming the basis for development of the design.

Stage 3D –
Design Development ...
covers developing in detail the approved concept, to finalise the design and definition criteria; establishing the detailed form, character, function and cost plan; defining all components in terms of overall size, typical detail, performance and outline specification;

and concludes with the Client approving the Design Development Report, setting out the integrated developed design for the Project and forming the basis for the development of production information.

Stage 3R –
Design Development ...
covers the review of the Stage 3D definition information prepared by others based on the Construction Contract Requirements (CCRs) for general conformity with design intent;

and concludes with Client acceptance of the Design Development Report.

Q: *Will all the tasks be allocated, throughout all the stages?*
A: *Not all tasks will be required on all projects; for example, under traditional procurement where the design team continues through the design development stage (Stage 4D) no one will review the development of the definition information during the design development and production information stages (Stages 3R and 4R) and these services will therefore not be required.*

Thus, Stages 3, 4 and 5 include both definition and review services. The early stages, Stages 1 and 2, only include definition services, and Stage 6 only covers review services (since by that stage the definition process is concluded).

Construction Industry Council

Stage 4D – **Production Information** …	covers the production of the final detailing, performance definition, specification, sizing and positing of all systems and components; and concludes with completion of integrated production information, enabling either construction (where the Contractor is able to build directly from the information prepared) or the production of manufacturing and installation information for construction.
Stage 4R – **Production Information** …	covers the review of the Stage 4D definition information prepared by others based on the Construction Contract Requirements (CCRs) for general conformity with design intent; and concludes with Client acceptance of the production information.
Stage 5D – **Manufacture, Installation and Construction Information** …	Covers defining the fabrication, manufacturing details and installation of all components, including all temporary works, connection details and builders' work; selecting proprietary equipment and components; and verification testing of components and systems and the creation of operation and maintenance manuals; and concludes with completion of manufacture, installation and construction information.	The definition process – defining what is to be constructed – does not end with completion of the production information and is not confined to the work undertaken by consultants; this is recognised in Stage 5D – where the tasks will be the responsibility of the contractor and its team (sub-contractors, specialists and suppliers); Stage 5D services will therefore usually be allocated to the contractor under the construction contract.
Stage 5R – **Manufacture, Installation and Construction Information** …	covers the review of the Stage 5D manufacture, installation and construction information prepared by others based on the production information for general conformity with design intent; and concludes with Client acceptance of the manufacture, installation and construction information.	*These services will be used, for example, in the case of traditional procurement, where the end-user client's consultants review the manufacture, installation and construction information prepared by the contractor. The consultants would therefore carry out Stages 1D, 2D, 3D, 4D and 5R, whilst the contractor will carry out Stage 5D.*
and Stage 6R – **Post Practical Completion** …	covers issues arising after Practical Completion; and concludes with completion of the Services.	NB Procurement planning approval, contract administration and construction monitoring are covered in separate tables.

CIC SERVICES HANDBOOK

Design and build

The CIC Services can be used with a **design and build** or **design, construct and operate contract**

If either a design and build or a design, construct and operate contract is used, and both the project client and contractor client use the CIC Scope of Services, the services will be fully integrated; the transfer of design responsibility from the project client's team to the contractor client's team can therefore be seamless and transparent.

Where two clients – eg project client and contractor client – use the CIC Scope of Services, the Tables will be used twice: once to allocate tasks to the project client's team, and again to allocate tasks to the contractor client's team. The first set of Tables will cover definition services prior to the construction contract being let, and review services (if used) thereafter. The second set of Tables will cover definition services after the construction contract is let. Planning approval, contract administration and construction monitoring services will be allocated as appropriate.

The **Construction Contract Requirements ...**

abbreviated to CCRs, are the provisions forming part of the Construction Contract setting out the Client's requirements for the development of the definition process.

For example, when tendering for a design and build contract (at the procurement stage), the CCRs are the employer's requirements for the contract is let, the CCRs are the employer's requirements modified by the accepted contractor's proposals – together with any other requirements of the building contract relating to the definition process.

Q: What if the client's consultants are novated to the contractor?
A: Effectively, the consultants provide their continuing services under a new contract, under which the contractor is the client – so one set of tables sets out the services pre-novation (with the project client as client) and another set the services post-novation (with the contractor as client).

NB when using DefinIT (the allocation software package), you can choose which symbols or letters you use.

Examples of the general allocation of roles for a design and build contract where the construction contract is let at the end of Stage 3D follow on the next pages.

In the first example (page 18) symbols are used to denote the developer Client and its Consultants (the Project Team). In the second example (page 19) letters are used to denote the design and build contractor Client and its Consultants and Specialists (the Construction Contract Team).

The third table (page 20) demonstrates the complete allocation of tasks by showing the two previous tables together.

CIC SERVICES HANDBOOK

Example of design and build: developer Client and its Consultants

Legend:
- ▨ inapplicable
- ⟿ to be allocated
- ◰ not used

GENERAL OBLIGATIONS	Client/Client Representative (= client)		Project Lead (= project manager)		Contract Admin (= architect)	Design Lead (= architect)		Architectural Design (= architect)		Civil & Structural Design (= structural engineer)		Building Services Design (= m&e engineer)		Cost Consultancy (= quantity surveyor)		Health & Safety Consultancy (= CDM coordinator)	
STAGES	Definition	Review	Definition	Review		Definition	Review	Definition	Review	Definition	Review	Definition	Review	Definition	Review	Definition	Review
STAGE 1 Preparation	●		●			●		●		●		●		●		●	
STAGE 2 Concept	●		●			●		●		●		●		●		●	
STAGE 3 Design Development	●		●			●		●		●		●		●		●	
STAGE 4 Production Information	⟿	●	⟿	●		⟿	●	⟿	●	⟿	●	⟿	●	⟿	●	⟿	●
STAGE 5 MI&C Information	⟿	●	⟿	●		⟿	●	⟿	●	⟿	●	⟿	●	⟿	●	⟿	●
STAGE 6 Post Practical Completion		●		●	●		●		●		●		●		●		●
Procurement	●	●	●	●		●	●	●	●	●	●	●	●	●	●	●	
Planning Approval																	
Contract Admin etc					●		●		●		●		●		●		●

Example of design and build: Contractor and its Consultants and Specialists

Legend:
- grey box = inapplicable
- / = not used
- C = contractor
- A = architect (Design Lead)
- Y = m&e specialist (Building Services Design)

STAGES		Client/Client Representative	Project Lead	Contract Admin	Design Lead	Architectural Design	Civil & Structural Design	Building Services Design	Cost Consultancy	Health & Safety Consultancy
GENERAL OBLIGATIONS		C (= contractor)	C (= contractor)	/	A (= architect)	A X	S Y	Y (= m&e specialist)	C (= contractor)	C (= contractor)
		Definition / Review	Definition / Review		Definition / Review	Definition / Review	Definition / Review	Definition / Review	Definition / Review	Definition / Review
STAGE 1 Preparation										
STAGE 2 Concept										
STAGE 3 Design Development		/	/		/	/	/	/	/	/
STAGE 4 Production Information		C (Def)	C (Def)		A (Def)	A (Def) / A (Rev)	S (Def)	Y (Def)	C (Def)	C (Def)
STAGE 5 MI&C Information		C (Def)	C (Def)		A (Def)	X (Def) / A (Rev)	Y (Def) / S (Rev)	Y (Def)	C (Def)	C (Def)
STAGE 6 Post Practical Completion						A (Rev)	S (Rev)	Y (Rev)	C (Rev)	
Procurement		C	C		C	C	C	C	C	C
Planning Approval		C	C		C	C	C	C	C	
Contract Admin etc		C	C		C	C	C	C	C	C

CIC/ServicesH
first edition 2007

Construction Industry Council

Example of design and build: Project Team and Construction Contract Team

Legend: ☐ inapplicable ⬜ not used

STAGES	Client/Client Representative		Project Lead		Contract Admin	Design Lead		Architectural Design		Civil & Structural Design		Building Services Design		Cost Consultancy		Health & Safety Consultancy	
GENERAL OBLIGATIONS	C		C			A		A / X		S / Y		Y		C		C	
	Definition	Review	Definition	Review		Definition	Review	Definition	Review	Definition	Review	Definition	Review	Definition	Review	Definition	Review
STAGE 1 Preparation																	
STAGE 2 Concept																	
STAGE 3 Design Development		/		/			/		/		/		/		/		/
STAGE 4 Production Information						A		A		S		Y		C			
STAGE 5 MI&C Information						A		X		Y	S	Y		C			
STAGE 6 Post Practical Completion									A		S		Y		C		
Procurement																	
Planning Approval																	
Contract Admin etc																	

CIC SERVICES HANDBOOK

The Client Brief and Project Brief

The Client Brief … is the outline statement setting out the Client's requirements and objectives at the outset of the Project;

It sets out the Client's wishes prior to the appointment of the Project Team.

the Project Brief … is the brief progressively developed from the Client Brief as part of the Services and agreed by the Project Team / Construction Contract Team and the Client from time to time.

It is the product of the work undertaken.

NB the term 'Project Team' is used where the Client is the project client, and 'Construction Contract Team' where the Client is the contractor.

The Project Definition … is the brief, design, programme and cost of the Project as agreed between the Client and the Project Team at any point in time. At the end of each stage it is fully integrated and formalised into the Project Brief (as set out below).

Q: Why so many different terms?
A: The different terms describe different manifestations of the agreed basis of the definition services. Each stage starts with a clear brief. During each stage the definition information (design and everything else) is developed through an iterative process which, by agreement, may modify the Project Definition. At the end of each stage the definition information is collected into a report for approval by the client (as set out above).

The Strategic Brief … is the Project Brief as approved by the Client at the conclusion of Stage 1D (Preparation);

it defines the Project objectives, business need, acceptance criteria and the Client's priorities and aspirations, and sets out the basis for the development of the concept definition for the Project;

As the definition process proceeds, at the conclusion of Stages 1D, 2D and 3D, the Project Definition is assembled into a report for approval by the Client:

• at the conclusion of Stage 1D this report is called the Strategic Brief …

the Concept Report … is the Project Brief as approved by the Client at the conclusion of Stage 2D (Concept);

it establishes the detailed brief, scope, scale, form and budget and sets out the integrated concept for the Project;

• at the conclusion of Stage 2D the report is called the Concept Report …

the Design Development Report … is the Project Brief as approved by the Client at the conclusion of Stage 3D (Design Development);

it develops in detail the approved concept to finalise the design and definition criteria and set outs the integrated developed design for the Project.

• and at the conclusion of Stage 3D the report is called the Design Development Report.

CIC SERVICES HANDBOOK

Construction Industry Council

Notes and terms used in the Tables

'Extent to be agreed'

A number of tasks are noted 'extent to be agreed', meaning that the parties to the contract should – before the contract is signed – agree the extent of work to be done.

With regard to Stage 6R services (post practical completion), it is good practice to agree how long the party undertaking the services will continue to work before regarding their services as having been completed (and requiring further fees or costs for more work).

In the case of Further Services, the extent of work to be done will always need to be agreed.

A list of tasks for which the extent is to be agreed can be found in the CIC Scope of Services.

Examples are the number of copies of drawings and documents to be provided (GO–17), the number of design options to be prepared and investigated (2D–33) and the extent of monitoring that construction of the Works is generally in accordance with the Construction Contract (CaCm–13).

CaCm–13. Construction monitoring **Extent to be agreed** (see FS–24 and 25)	Rece... whether of W... in ac... with... Cont...

cross referencing

In some rows, related provisions are cross referenced.

For example, row 2D–19 covers establishing the need for construction or other specialist advice, and there is cross reference to the procurement services which cover the selection and appointment of specialists (Pr–9 to 13).

'Applies if tasks to be undertaken by someone who did not undertake relevant tasks in Stage 3D'

This and similar notes indicate that the row may or may not be applicable, depending upon whether the tasks are to be undertaken by Consultants, Specialists or Contractor developing definition work prepared by others or not.

Alternatives (applicable depending upon the circumstances) are included, for example Design Development Report or CCRs (Construction Contract Requirements) and Project Team or Construction Contract Team.

For example, row 4D–2 covers consideration of the Design Development Report / CCRs.

If, say, architectural design is to be undertaken by an architect who also undertook Stage 3D tasks, this task does not apply.

> With Project Team, consider Design Development Report and advise on issues requiring supplementary information or clarification.

If, however, building services design is to be undertaken by a specialist taking over from a consulting engineer, the task will apply.

> With Construction Contract Team, consider CCRs and advise on issues requiring supplementary information or clarification.

If at the beginning of Stage 4D a design and build contractor has been appointed, and the tasks are to be undertaken by the contractor's team, the whole row will apply.

4D–2. Design Development Report / CCRs – consideration **Applies if tasks to be undertaken by someone who did not undertake relevant tasks in Stage 3D**	Resu... Team... Cont... consi... Deve... CCR... on iss... supp... inforr... clarifi... instr...

3D–35. Interim report Optional

part way through the design development stage, there is provision for an interim report, which some clients may require.

This marks the division between RIBA revised Stages D and E.

CIC SERVICES HANDBOOK

'PL prepares tender'

Project Lead	A
Prepare tender documents, including preliminaries and pricing documents. Finalise selection criteria. Report to Client, obtain approval and issue. ***alternative to 21B (PL prepares tender)***	A o A s a (l
Monitor preparation of tender documents and	A o

In Procurement Services, provision is made for either the person undertaking the Project Lead role (PL) or the Cost Consultancy role (CC) to prepare the tender documents and undertake related tasks.

The alternative rows are denoted 'A' (PL prepares tender) and 'B' (CC prepares tender); either 'A' rows should be chosen or 'B' rows.

(PL prepares tender)	A
Monitor preparation of tender documents and finalisation of selection criteria. ***alternative to 21A (CC prepares tender)***	A o A s a (l

Pr–32 to 40 optional

These Procurement Services, covering second stage tenders, are marked as optional.

procedures

The Tables are drafted on the basis that communication with the Client is through the Project Lead; the Design Lead receives advice from the designers (Architectural, Civil & Structural and Building Services), co-ordinates their input and advises the Project Lead, who then reports to the Client and where appropriate obtains instructions.

Other arrangements can always be agreed between the parties.

Management Procedures

At the beginning of each Stage, those undertaking the Services are to agree and implement Management Procedures;

these are defined as management systems, decision-making processes, reporting procedures and information exchange standards to facilitate co-operation and communication and to enable and monitor:

- progress of the production or review of the Project Definition
- adherence to programme, cost and contract requirements and
- integration of the Project Definition across roles.

Also included are procedures for dispute avoidance and the early resolution of disputes.

Management Procedures are widely defined, giving the parties the opportunity to put in place whatever procedures they require.

The Specific Scope Schedule

The **Specific Scope Schedule** ... allows the parties to allocate the primary responsibility for the definition of components.

So the potential for confusion over responsibility for specialist pre-cast concrete design, underground drainage, sprinkler systems or external works should be a thing of the past as the elements and stages for which design services are required are defined. It also provides a clear framework for change management.

For example, an architect may have primary design responsibility for architectural components generally in Stages 2D, 3D and 4D, but responsibility for the façade (item 4.12) may be taken by a specialist designer in Stages 3D and 4D;

SPECIFIC SCOPE SCHEDULE

Schedule of primary definition responsibility
The allocation of primary responsibility for the definition of components

Stage	2D	3D	3R	4D	4R	5D	5R	Notes
1 General								
1.1 Site layout								
1.2 Setting out: dimensions, levels, falls								
1.3 Temporary works								
1.4 Cost estimating								
1.5 CDM (design risk assessments)								
1.6 Demolition and site clearance								

A yet further level of detail can be found in other industry documents, for example *A Design Framework for Building Services* published by BSRIA and *Allocation of Design Responsibilities in Constructional Steelwork* published by BCSA.

5 Structure

5.1	Retaining walls – concrete part of building								
5.2	Retaining walls – masonry part of building								
5.3	Ground bearing slabs, structural fill, blinding								
5.4	Structural steelwork								
5.5	Steelwork connections								
5.6	Steelwork corrosion protection								
5.7	In situ reinforced concrete frame								
5.8	Reinforcement drawings and bending schedules								
5.9	Precast structural concrete								

Components and elements of the project are listed under the following headings:

- General
- Third party liaison
- Surveys / existing structures
- Architectural
- Structure
- Building services
- Builder's work
- Miscellaneous
- Drainage
- External works and landscaping and
- External liaison / off site works.

The parties may add further components or elements, as required.